Cool Careers in MATH

Contents

Tony

Troy

Karl

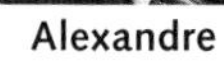

Alexandre

Daphne

Ermelinda

Trachette

Jonathan

Victoria

Deborah

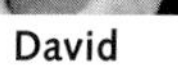
David

Nancy

What Do You Want to Be?

Is working with numbers one of your goals?

The good news is that there are many different paths leading there. The people who use math in their work come from many different backgrounds. They include mathematicians, biologists, chemists, physicists, computer scientists, engineers, teachers, and many more.

It's never too soon to think about what you want to be. You probably have lots of things that you like to do—maybe you like doing experiments or drawing pictures. Or maybe you like working with numbers or writing stories.

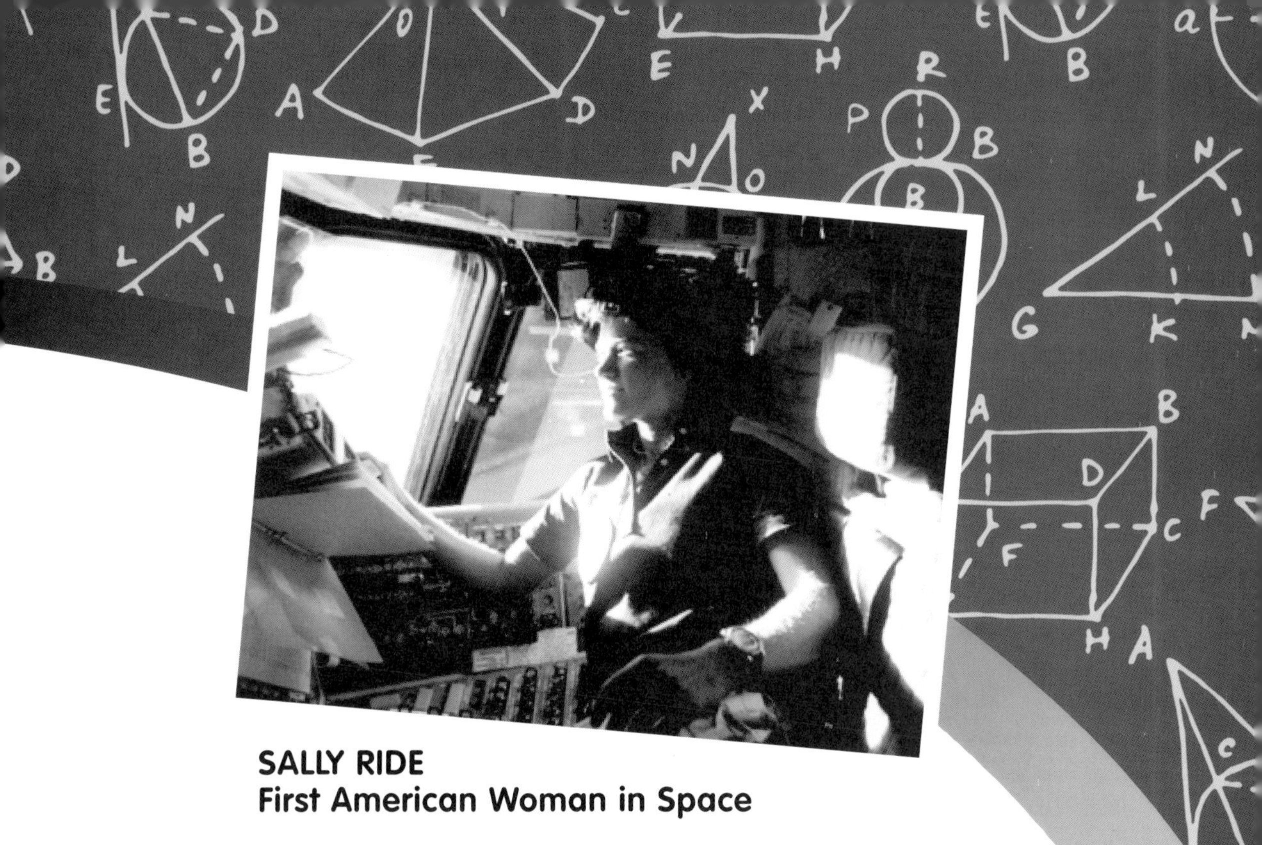

SALLY RIDE
First American Woman in Space

The women and men you're about to meet found their careers by doing what they love. As you read this book and do the activities, think about what you like doing. Then follow your interests, and see where they take you. You just might find your career, too.

Reach for the stars!

Sally K Ride

TONY DEROSE

Pixar Animation Studios

Tony's favorite Pixar character? "Bob Parr—Mr. Incredible. I really identify with him."

Mathematical Movies

When you saw *WALL-E*, *Monsters, Inc.*, *Cars*, *Up*, or any other cool animated movies, did you know math made them possible? Just ask Tony DeRose at Pixar, the film studio behind those movies. Tony isn't an animator—he doesn't even draw well. Instead, Tony creates the software tools that Pixar's computer animators use. "Since we make our movies on computers, everything gets reduced to numbers," Tony says.

No Quick Draws

One big challenge for animators is portraying the complex geometry—shapes—of hair, clothing, and skin as they move. Imagine someone talking. Animating *just* the character's face—wrinkles, eyebrows, and all—is one big math problem! Luckily, Tony created software to speed up the process. The software tracks about 3,000 points on the face. "For each frame in the film, we have to determine where each of those points is. And those points have to move so they look like a person," Tony says. "That's a lot of information—way too much for a person to produce by hand." Even with the help of Tony's software, it can take four years to make a movie!

Digital Geppetto

Tony has always loved discovering new ways to create things. So what if Tony's first try to build a robot out of old toys wasn't quite a success—he was only four! It still inspired him to continue tinkering. Look where that tinkering took him!

Tony captures real life on film while vacationing in the Galapagos Islands.

An animation scientist . . .

develops the technology needed to make animated movies, television shows, and video games. Tony develops software that animators use to create realistic-looking animated characters. Other **animation scientists**

- model how water moves.
- research how light falls on moving objects.
- study facial movements to make characters more lifelike.

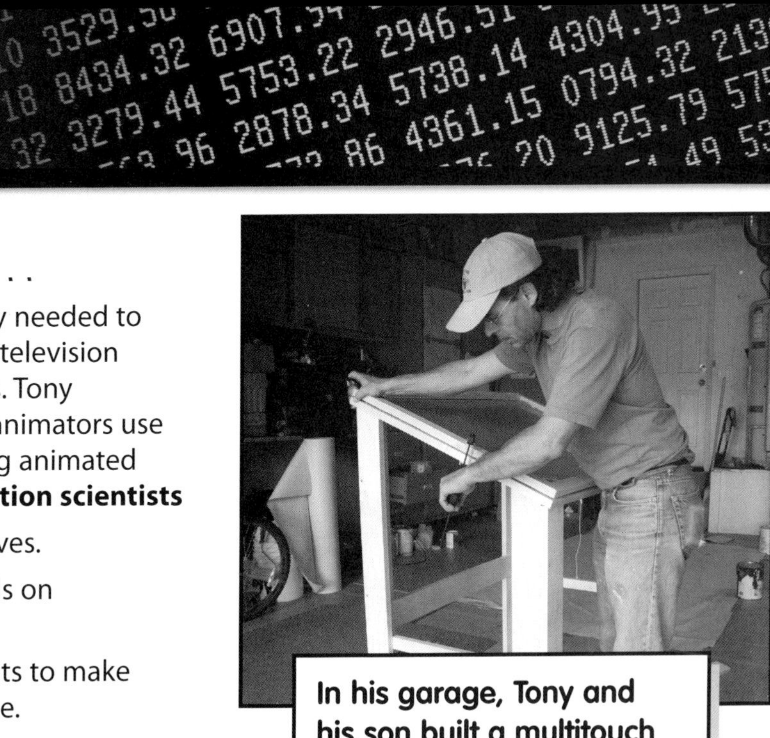

In his garage, Tony and his son built a multitouch computer workstation.

Flip It

Create your own animated movie by making a flip book. Here's how.

- Start with about 25 index cards without lines.
- Think of a simple story to animate, such as a seedling growing into a plant or a dog fetching a bone.
- Draw one scene on each card. The main parts that are supposed to move, such as the seedling or dog, should get bigger or move by $\frac{1}{8}$ to $\frac{3}{8}$ inches from card to card.
- When you're finished, clip the cards together with a binder clip.

Flip the pages and enjoy the show. Trade flip books with your classmates.

Movie-Magic Math

Animated movies are made from a series of drawings. They flash by on screen at 24 frames—drawings—per second. The running time of *Cars* is 117 minutes. How many frames make up the movie?

The Gap

During a movie, one still picture after another flashes before your eyes. But why do you see movement instead of still pictures? It's because of an illusion called "apparent motion." Your brain fills in the gaps between each of those still pictures to create the illusion of smooth motion.

Check out your answers on page 36.

TROY MIYASATO

Ferraro Choi and Associates, Ltd.

It's a Math-terpiece!

What do you think of when you look at the Empire State Building or Space Needle? Math? Probably not. Well, think again. According to architect Troy Miyasato, "Math comes in from almost the very start when you're designing a building." He uses volume and area to calculate the overall size of the building and its rooms. There's geometry behind all the building's shapes and angles. In addition, computer models use math to analyze things such as air flow through the building and glare from the Sun at different times of the day.

More Than a Building

Troy's firm specializes in "green" architecture—designing buildings that incorporate Earth-friendly, sustainable concepts. Their design for the Hawaii Gateway Energy Center earned the highest rating for energy efficiency and environmental design. With elements like solar panels and a cooling system that uses nearby cold, deep seawater, the center generates more energy than it uses. Now that's cool.

The Art of Math

Growing up, Troy was hardly a math nerd. He loved the arts and took classes in sculpting, metalwork, and drawing. He learned to appreciate math as a tool that could help him lay out on paper the beautiful designs in his head. Later, in college, when Troy took higher level math, he says he came to view the application of math as an art in itself.

Taking care of our planet for his girls—and all future generations—is one of Troy's inspirations for green design.

An architect . . .

designs buildings and other structures, then oversees their construction. Troy uses his art and math skills to design new buildings and remodel and upgrade existing buildings, especially hospitals and schools around Hawaii. Other **architects**

- plan new housing communities.
- remodel and upgrade existing buildings to meet environmental standards.
- design bridges, airports, and other structures used by millions of people.

Up a Tree

You've been asked to design the neighborhood tree house—and it has to be environmentally friendly. Brainstorm a design with a team.

- Will your tree house have a platform? Sides? A roof?
- What materials will you use? Recycled plastic or wood?
- How will it protect everyone from rain and wind?
- What is the scale of your design? For example, 10 centimeters could equal 1 meter.

Draw your design. Teams take turns sharing their ideas with the class. Then combine the best features for the coolest, greenest tree house.

It's Symbolic

With a partner, create a floor plan for any room in your school. Take measurements, then make a drawing to scale. Fill in details, but don't label anything with words—*use only symbols*. Here are some samples of symbols that architects use as labels, but you'll need to create more.

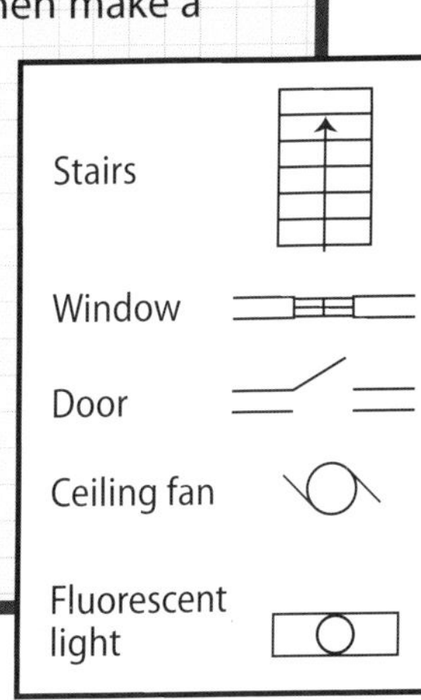

Geom-pet-ry

The Dog Circle Day Care hires you to design its new building—as a circle, of course. Here are their specifications.

- A dog run—going through the center of the building, from the north door to the south door
- Two play areas—one starts at each door and extends 16 meters to the fence at the edge of the property
- The property is square—65 meters by 65 meters

Sketch it out. How large, in meters, will you be able to make the circumference of the building?

Check out your answers on page 36.

Karl Schaffer

Dr. Schaffer and Mr. Stern Dance Ensemble

Puzzling Moves

When Karl Schaffer was young, he never came down the stairs the same way twice. Maybe Karl would hop or jump down two steps—and then hop back up one. He liked to puzzle out a pattern—and then throw his whole body into recreating it. "In a way, that's what I am still doing—physical activities that have a mathematical component to them," Karl says.

Rumbas and Rhombuses

While earning his degree in math, Karl started studying dance. "For the first year, I was terrible—I literally had trouble putting one step in front of the other." Karl didn't stumble for long. Today he creates and performs dances professionally. He also teaches math at De Anza College. So which is Karl's hobby and which is his job? Neither—and both. Karl might use rhythms in dance to explain common denominators in math—or the other way around. That can mean tap-dancing out a four-beat pattern to six-beat music. Repeat it three times and a pattern emerges around the least common multiple—12. Feel the beat and do the math.

Dance + Math = Learning

Karl brings dance to his math classroom and math to the stage. "There are connections between them—they are not always obvious. So my work is making them visible." Karl also helps teachers use dance to make math class more interesting. "Physical activity is a way of being engaged in learning."

Karl choreographed this dance called "Grasp Bird's Tail"—a phrase taken from a Tai Chi movement.

A choreographer . . .

creates and arranges dances. Karl uses math to create new dances and to better understand classic dances. Other **choreographers**

- create routines for synchronized swimmers.
- train dancers for music videos.
- judge professional dance competitions.
- create dances for ballet troupes.

Math Moves

Do you know how to do a hip-hop toe drag? How about a shuffle-hop-step like a tap dancer? List as many different types of dance as you can. Check off those where you know at least one step. Now research one or two steps of a dance you've never tried before. Then

- write step-by-step instructions that tell exactly how to do the moves.
- use simple drawings to illustrate your instructions.
- demonstrate to your class what you've learned.
- ask volunteers—and perhaps your teacher!—to try the steps with you.

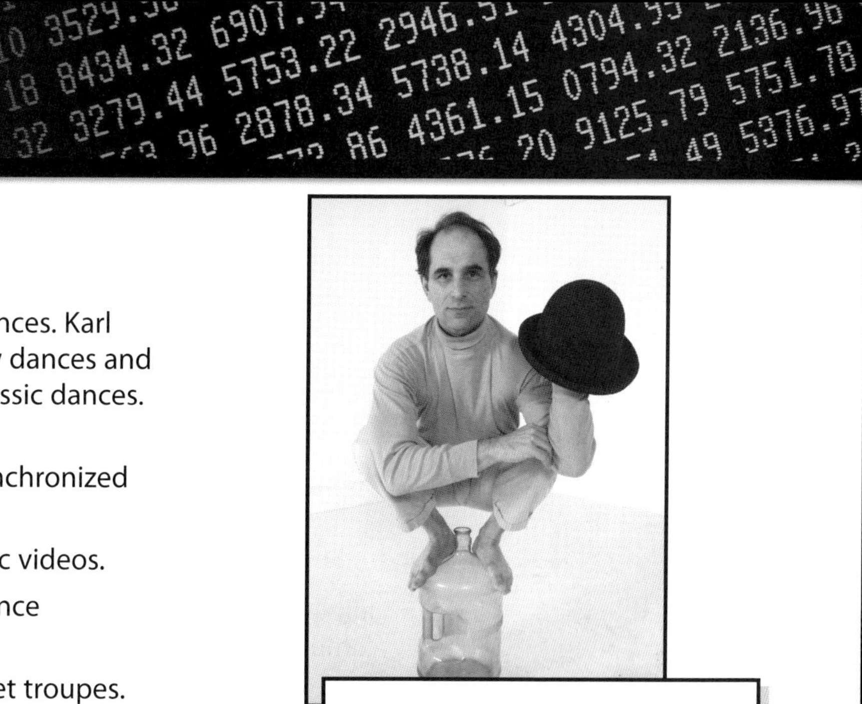

In one dance, Karl had dancers twirling water bottles—like the one he's perched on—while dancing!

Name Game

Karl and his colleague Erik Stern have their students sound out the letters of their first name by clapping their hands for a vowel and slapping their thighs for a consonant. For example, the name K-A-R-L would be clap-slap-clap-clap.

What pattern represents your name? Practice your name pattern so you can do it without pausing. Play your name for your classmates.

Now replace the slaps and claps with dance movements and dance your name. Present your name to the class.

Civil Systems Engineer

Alexandre Bayen

University of California, Berkeley

Math to the Rescue

There's good news for busy air traffic controllers and pilots. Alexandre Bayen wants to make their jobs easier. This civil systems engineer is using his math and engineering skills to turn the problem of air traffic congestion into a math problem. He believes that solving it could lead to a warning system that will not only prevent crashes, but will also cut down on airport delays. "Mathematics is a fantastic tool," he says.

Faster, Safer, Better

To create computer tools that help air traffic controllers, Alexandre is coming up with math equations for all the possible flight paths that could lead to a collision between two planes. The information could be programmed into a control center. There it would automatically warn air traffic controllers when planes are in danger. The same system could also make airports more efficient. It would inform air traffic controllers when it's safe to line up airborne planes for landing. "It's nice to be able to contribute to making things work better," Alexandre says. That adds up.

"Engineering is really about modeling everyday problems and putting equations behind what you see."

In this control center, real people react to simulations as they test future technology.

A civil systems engineer . . .

uses math and engineering to analyze complicated systems that are made up of many parts, each affecting the others. Alexandre is working to make air traffic flow better at airports. Other **civil systems engineers**

- improve the flow of highway traffic.
- use electronics to monitor earthquakes.
- work on improving unpiloted airplanes.

Touchdown Time

In your job as an air traffic controller at Earhart Airport, you're developing a landing schedule for three incoming flights. Check the details below. Then calculate the landing time for each flight.

Flight	Flight Distance (km)	Speed (kph)	Takeoff (Earhart time)
001	1,800	300	10 A.M.
002	2,100	350	9 A.M.
003	1,200	400	11 A.M.

Plane Funny

Q. Did you hear about the student pilot who was taking her flight test and flew through a rainbow?

A. She passed with flying colors!

Skill Drill

What skills do you need to be a civil systems engineer? According to job recruiters, here are some important ones.

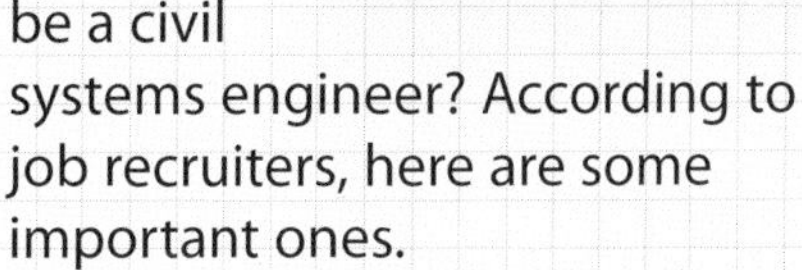

Math—Use arithmetic, algebra, geometry, calculus, and statistics.
Critical Thinking—Use logic and reasoning.
Science—Use scientific process to solve problems.
Active Listening—Ask questions and pay attention to what other people are saying.
Active Learning—Understand new on-the-job information for problem solving and decision-making.
Problem Solving—Identify problems, develop and evaluate options, and implement solutions.

With a partner, take turns playing the role of a job recruiter and job candidate. Interview each other and take notes as you find out which skills you have or are learning, and how and when you use them.

Check out your answers on page 36.

DAPHNE KOLLER

Stanford University

Programming Phenomenon

What pops into your head when you hear "computer programmer"? People staring at monitors all day? Daphne Koller hopes you hit "delete" on that image. "That's really not what computer science is all about," she says. Daphne builds mathematical computer models to better understand real-world phenomena. One of Daphne's projects is to create a computer model of how a *single* human cell works. Sound simple? Hardly. Cells, such as these red blood cells, are incredibly complex! They're constantly sending and receiving signals, and converting and releasing energy.

Micro-processing

Biologists can measure various aspects of the activity inside cells. But these are just numbers. Daphne uses those numbers to figure out what's really happening. She constructs math equations that could be used to describe a cell. And she gives the computer different options for those equations. The computer searches to find the equations that best fit the biologists' measurements. Since our knowledge is incomplete, Daphne's model incorporates probability. It might know that if an observation is based on a single measurement, it's probably not 100 percent accurate. With more observations, the probabilities improve, and so does the model. Why all the fuss over a tiny cell? To understand the whole body, you need to understand its building blocks—tiny cells.

Cool Code

Until she was 12, Daphne lived in Israel. Then she came to the U.S. with her parents. Here, she joined her school's computer club and learned how to program on a dinky computer. "It was remarkable that you could write a few lines of code and get a computer to do all sorts of cool things," Daphne says.

A computer scientist . . .

comes up with and improves ways to use computers to solve problems. Daphne builds mathematical models on computers to uncover new truths about the physical world. Other **computer scientists**

- figure out new ways to protect private information.
- program robots to analyze situations and behave more like humans.
- devise better ways to transmit information between computers.

Daphne has traveled all over the world with her husband and two children—hiking, sailing, and scuba diving.

Can You Cell It?

Daphne uses computers to model how a cell works. Team up with your classmates to create a model of what a cell looks like.

- Find a labeled picture of an animal cell.
- Decide whether you want to make a 2-D or 3-D model.
- Your model should include these cell parts—cell membrane, cytoplasm, mitochondria, nucleus, lysosome, Golgi apparatus, and endoplasmic reticulum.

Be creative and have fun with the materials you choose—a partially inflated balloon could be a nucleus. Make a cell fact card that tells what each cell part does, and display it with your model.

Odds on Dimples

Like most traits, having dimples depends on the genes you inherit from your parents. Dimples are a dominant trait. That means you need to inherit only one gene for dimples (D) to have dimples. If you inherit a recessive gene for no dimples (d) from each of your parents, then you won't have dimples. See for yourself.

This Punnett square shows the dimple trait for two parents.

	♀ D	♀ d
♂ d		
♂ d		

- Copy and complete the square.
- What are the odds of a child in this family having dimples?
- Give your answer as a fraction, a percent, and a decimal.

Check your answers with those of your classmates.

Check out your answers on page 36.

ERMELINDA DELAVIÑA

University of Houston-Downtown

Ermelinda encourages her students to both ask and answer questions during class.

Teaching the Next Generation

Ermelinda DeLaViña grew up in a single-parent home in a poor neighborhood. She never imagined herself going to college. But her ninth-grade algebra teacher encouraged her to believe in herself. Now that Ermelinda is a mathematics professor, she's passing on the favor. She loves helping her students aim for the stars.

Connecting the Dots

Ermelinda also loves doing math research. She works on complex problems in a branch of math called graph theory. But the graphs she works with are different from those you usually see. These graphs are networks of dots with lines connecting some of them. They help her visualize relationships within a problem. For example, imagine a teacher wants to randomly split students into study groups. How many groups will it take to make sure that none includes a pair of chatty best friends? Ermelinda would use dots to represent the students and lines to indicate which students are best friends. She could color the dots different shades to indicate different groups. By sketching scenarios like these, Ermelinda can grasp tricky math problems much more easily.

Good Graffiti

Ermelinda and a colleague developed a computer program, called Graffiti, which comes up with interesting math problems. It sorts through a database of millions of graphs, looking for possible relationships between them. Then it's up to mathematicians to prove whether those relationships are valid.

A math professor . . .

studies an area of math and teaches math to students. Ermelinda teaches math courses, from algebra and calculus to higher-level classes, while conducting research on graph theory. **Math professors** also

- mentor and advise students on career paths.
- work with individual students on their own research in math.
- review scholarship applications.
- attend conferences to meet and collaborate with other mathematicians.

Behind Ermelinda is an exercise ramp she built for her dog Indie.

Colorful Proof

A conjecture is a mathematical statement that is thought to be true. Find out for yourself if the four-color conjecture is true.

- The conjecture states that you can color regions of a map, such as states, using only four colors in a way that no two states that share *a full border*, not just a corner, have the same color.
- Work with a partner and test the conjecture.
- Start with a black and white map of the *continental* U.S.

Sound easy? Start coloring!

Ancient Truths

One of the first mathematicians, Pythagoras, lived from 560 to 480 B.C. He discovered some interesting things about triangles. Make this drawing and see what you discover.

- Draw a triangle with a right angle—90°.
- Now draw three rectangles of the same height, using the sides of the triangle as the base of each rectangle.
- Calculate and compare the areas of the three rectangles.

What did you discover?

A Friend Indeed

Write an informative entry in your About Me Journal that describes how a teacher, coach, or other adult has helped you believe in yourself. Then share your entry with a classmate.

TRACHETTE JACKSON

University of Michigan

"Mathematics is not just a surface subject. Learn it deeply, and then you can use it to do anything in the world."

Fuzzy Math

In college, Trachette Jackson heard that a math researcher was giving a talk about how the leopard got its spots. She was curious. What did that have to do with math? Amazingly, she learned that math could be used to predict the pattern of spots on a newborn leopard. It was the first time Trachette saw how math could be applied to a science like biology. "It was an eye-opener for me," she says.

Equations for a Cure

Today, Trachette uses math to look at possible treatments for cancer. She makes mathematical models—series of equations inserted into a computer program. Her models explain how cancer cells grow, divide, and in some cases migrate from one place to another. Once she has a mathematical way of describing cancer growth, she can run simulations to see how new treatments might affect it. Trachette works closely with cancer researchers who use information from her models to improve their experiments in the lab.

Teaching Teachers

One summer in high school, Trachette attended a math camp for girls and minorities. When a teacher encouraged her to stick with math, she listened. Now Trachette is a professor herself, teaching the next generation of mathematicians.

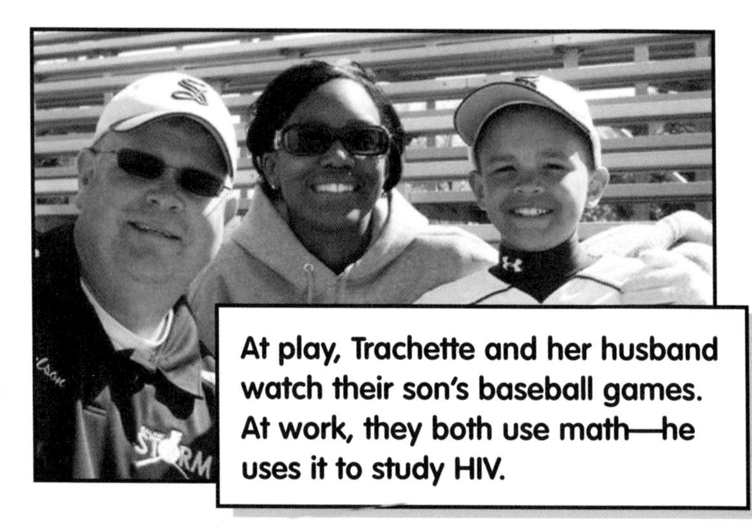

At play, Trachette and her husband watch their son's baseball games. At work, they both use math—he uses it to study HIV.

A mathematical biologist . . .

applies math to unsolved problems in the life sciences. Trachette helps biologists evaluate new cancer treatment strategies. Other **mathematical biologists**

- study how HIV evolves to resist treatment with drugs.
- build computer models to probe how regions of the brain work.
- simulate how plant and animal populations respond to environmental change.

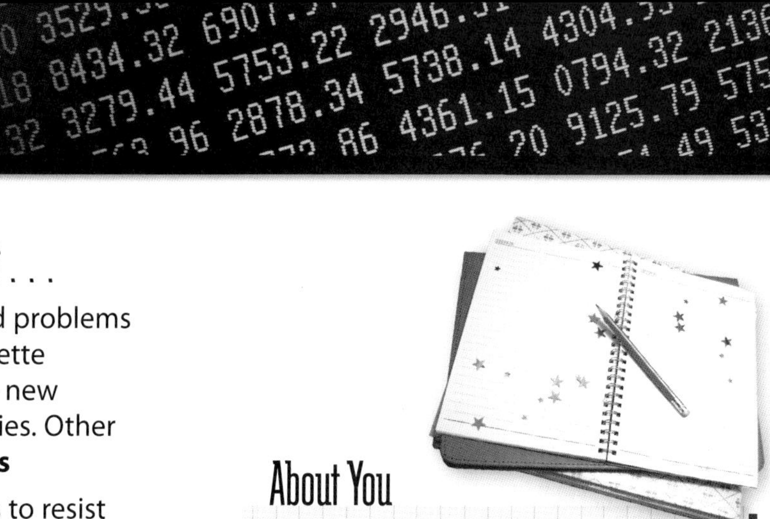

About You

Trachette participates in a program where she is a mentor to students. What's a subject you want to know more about? How could you benefit from having a mentor? Write responses to both questions in your About Me Journal.

Why Cells R Little

The cells that make up most plants and animals are tiny—between 10-30 micrometers in diameter. Why? It's because of the relationship between volume and surface area. See for yourself.

- Say you have one 4-centimeter cube, eight 2-centimeter cubes, and sixty-four 1-centimeter cubes, all with the same *total volume*—64 cubic centimeters.
- Calculate the *total surface area* for each set of cubes.
- Then figure out the ratio of the *total surface area* to *total volume*.

As a class, discuss your findings and why it's vital for cells to be little.

Growing Numbers

Suppose a certain cell divides into 2 cells every 5 minutes. If each cell divides at a constant rate, you can predict how many cells there would be after 1 hour. Here's how.

- First find out how many 5-minute intervals there are in 1 hour. That number (n) is the exponent by which 2 is raised, or 2^n. In this case, 2 is the number of cells that result after a single 5-minute interval.
- Now calculate how many cells there would be after 1 hour. Do the calculations by hand or use a calculator.

You may be surprised by the answer!

Check out your answers on page 36.

JONATHAN FARLEY

Hollywood Math and Science Film Consulting

"I love math. It's not just something I do to get a paycheck. I would do it for free."

Fill in the Blank

A career questionnaire Jonathan took in his tenth grade English class told him that he was well suited to become a mathematician. That's when he decided to become a mathematics professor.

Math, Hollywood Style

Jonathan Farley wears many hats as a mathematician. He does research and has taught at universities around the world. He also co-founded a company that helps keep Hollywood scripts honest . . . at least their math! In an episode of the TV show *Medium*, the main character hears the voice of a mathematician talking about complex math. It's an actor's voice, but it's real math. Jonathan gave the lines to writers who wanted to make the character sound like the real deal.

Mysteries and Solutions

Jonathan does cutting-edge research in which he looks for solutions to math problems no one has ever solved. Working in a branch of abstract algebra, he's answered questions that had remained mysteries for decades! But you don't have to be at his level to think about unanswered math questions. For example, a perfect number is one that equals the sum of its proper divisors—those whole numbers that divide evenly into the original number. So 28 is a perfect number because 1 + 2 + 4 + 7 + 14 = 28. The peculiar thing is that no one knows of a single odd perfect number. "No one knows if there is one or if there isn't one," Jonathan says. "And every mathematician in the world would love to know the answer."

A mathematician . . .

solves problems using numbers, symbols, equations, theories, techniques, and logical reasoning. Jonathan tries to answer unsolved problems and gives math advice to TV and film producers. Other **mathematicians**

- use measurements and statistics to study environmental problems such as air and water pollution.
- calculate flight plans for airplanes and spacecraft.
- find ways to secure secret information for companies or for the government.

Movie Math

In 1978, the average price of a movie ticket was $2.34! By 2008, it was $7.18. Use your math smarts to predict what a movie ticket will cost in the future.

Year	Price
1978	$2.34
1988	$4.11
1998	$4.69
2008	$7.18

- What was the average yearly change in the price of a movie ticket between 1978 and 2008?
- Now predict what the average price of a movie ticket might be in 2018.

Compare and discuss your calculations with a classmate.

A Penny Saved Is a Penny Learned

Jonathan uses diagrams to solve problems that have been mysteries for decades. Follow Jonathan's lead and solve this mystery.

The diameter of a penny is 1.9 centimeters. How many pennies would it take to line them up from wall to wall on a school gym floor that is 15 meters wide? How much money does that equal? Once you've figured it out with pennies, try it again with quarters.

The Winning Formulas

Here are some formulas, properties, and constants that mathematicians use. You've probably used them, too. Write down what they are. If you're not sure, look them up. Check your answers with a partner.

1. $A = \frac{1}{2} bh$
2. $V = \frac{1}{3} \pi r^2 h$
3. $C = \pi \times d$
4. $a^2 + b^2 = c^2$
5. $a(b + c) = ab + ac$
6. $a + b = b + a$
7. π
8. $°F = \frac{9}{5} °C + 32$

Check out your answers on page 36.

VICTORIA COVERSTONE

University of Illinois at Urbana-Champaign

Mission Critical

Imagine an asteroid is heading straight for Earth. Unstopped, it will hit us. We have ten years to prepare. What should we do? What's the asteroid made of? Can we blow it up? Can we redirect it? That's what Victoria Coverstone asked her college students last year. They had to design spacecraft, instruments, and strategies that could save our planet.

The Future of Flight

In her own research, Victoria programs computers to figure out the best paths for spacecraft to take on long trips. And she specializes in the paths of future "low-thrust" spacecraft, such as those that use solar sails, which sunlight alone will push. That requires lots of difficult math, because these spacecraft will need to adjust their paths constantly along the journey.

Fresh Eyes

Victoria is fascinated by her fancy equations, but she says she most enjoys working with her students. "They're bright, they're full of energy, and they're passionate about where they see our society could go," she says. "It keeps you fresh."

A mission designer . . .

plans a space mission that can include one or more spacecraft. Victoria designs flight paths for low-thrust spacecraft. Other **mission designers**

- select destinations.
- design spacecraft.
- choose instruments and experiments to put in spacecraft.

Sail Away

Someday, solar sails—gigantic mirrors—may propel spacecraft through our solar system and beyond. As sunlight strikes the reflective sails, it bounces off them. This gently pushes the spacecraft through space—much like wind pushes a sailboat across water. Solar sails could help spacecraft reach speeds of 30,000 kilometers (18,641 miles) per second!

At that speed, how long would it take a spacecraft to reach these planets and star? Round your answers to the nearest whole unit.

Body	Distance from Earth
Venus	40 million kilometers
Neptune	4.3 billion kilometers
Alpha Centauri	71 trillion kilometers

Hello Pluto

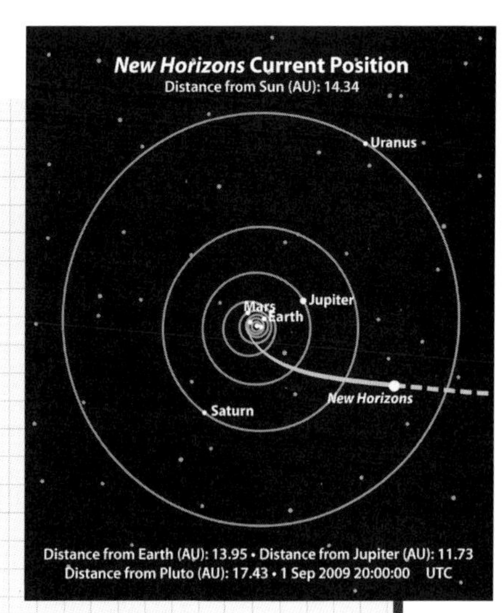

A mission designer has the awesome job of calculating the best paths for spacecraft to travel. Here's the path of *New Horizons*—the robotic spacecraft heading to Pluto, which it should reach in 2015. This image shows *New Horizons'* position along its trajectory or path.

- The solid line shows where *New Horizons* has traveled since it launched in 2006.
- The dotted line shows the spacecraft's future path.
- On September 1, 2009, *New Horizons'* distance from Earth was 13.95 AU.
- As you probably know, one AU, or Astronomical Unit, is the average distance between the Sun and Earth, or about 149.6 million kilometers (93 million miles).

On September 1, 2009, how far from Earth was *New Horizons* in kilometers? In miles?

Check out your answers on page 36.

Deborah Estrin

Center for Embedded Networked Sensing
University of California, Los Angeles

Deborah holds up the brains of the networks she designs—a sensor node.

Sensing Change

The rain forest is loaded with life—trees, vines, and other exotic plants; insects, birds, and other native animals. Years ago, while on vacation in Costa Rica, Deborah Estrin was struck by the complexity of the tropical ecosystem. She wondered if a web of sensors set up throughout the forest could help scientists understand it better. So, she got to work. Using mathematical logic and computer programming, Deborah started designing networks of wireless sensors. And yes—she even helped set up a network of sensors that collect and analyze data in a Costa Rican rain forest! The data help scientists understand all sorts of things, such as how climate change is affecting various plant species and the nesting habits of birds.

Pocket Sensors

Closer to home, Deborah and her team are using mobile phones as sensors. Using their software, a smart phone can upload your location to a database several times a minute. Then the system uses clever statistics and other data sources to estimate whether you are walking, running, or driving. It can also estimate your personal carbon footprint and exposure to air pollution. From the individual to the global, wireless networks are changing what we know about our world.

Seeds of Innovation

With computer science professors for parents, Deborah grew up around talk of innovation. "I've wanted to invent things for as long as I can remember," she says. Today, her innovations are breaking new ground.

Dinner in the rainforest. A hummingbird aims for the sweet nectar in the center of a passion flower.

A networks expert . . .

uses computer programming and math to link machines together efficiently. Deborah uses networks of sensors, computer science, and math to collect and process data about the physical world. Other **networks experts**

- work to increase the speed with which computers communicate.
- design new methods to protect private online information.
- develop systems to link new technologies.

Tiny Brains

One way to measure the power—or smarts—of a microchip is by counting the transistors packed onto its surface. Forty years ago, microchips contained about 2,000 transistors. Since then, that number has doubled about every two years.

1. Given that rate, how many times would the number double?
2. If the number has doubled that many times, how many transistors would you expect a microchip to contain today?
3. How many times more powerful are today's microchips compared to those made 40 years ago?

Waka Wake-up Call

Scientists often use sensor networks to track and monitor wildlife. Deborah once used networks of microphones to locate acorn woodpeckers by listening for their distinctive "waka-waka-waka" call. Think of a wild animal that lives in or around your community. Make a science poster showing how and why you would use a network of cameras, microphones, or other sensors to study it.

It Makes Sense

Deborah and her team have placed a network of real-time cameras and sensors in the San Jacinto Mountains of California. Robotic cameras watch bluebird nests, motion sensors detect predators, and environmental sensors monitor soil chemistry. Eventually there will be 100 sensors covering about 30 acres of wooded areas.

If 1 acre equals about 4,047 square meters (43,560 square feet), how many square meters will the sensors cover? How many square meters will *each* sensor cover? Now, convert each distance to feet.

Check out your answers on page 36.

DAVID ANNIS

SportsQuant.com

Numbers Game

For someone who loves numbers and sports as much as David Annis does, sports statistics really count. Statistics is the mathematics of collecting and analyzing numerical data to draw conclusions and make predictions. Luckily for David, sports are full of numbers. Runs, errors, and steals add up in a baseball game. Touchdowns, turnovers, and conversions pile up in football games.

Three Points, Two Opinions

Have you ever questioned a coach's play? David sure has—that's because he does the math before the coach even runs a play! Imagine a professional basketball team with a three-point lead. Should they foul their opponent before they can attempt a three-point shot—or step up their defense? David's analysis suggests the team should go for the foul. Do you have a favorite basketball team? Be sure to pass on that stat to the coach.

Sports Stats

Throughout school, David's engineer dad quizzed him with challenging math questions. "I always enjoyed solving math problems that I hadn't seen before," he says. Now David has his own Web site where sports fans, coaches, players, and sports writers check out his sports statistics. That adds up to a win for a math whiz and sports enthusiast like David, "I get a lot of personal satisfaction out of doing this."

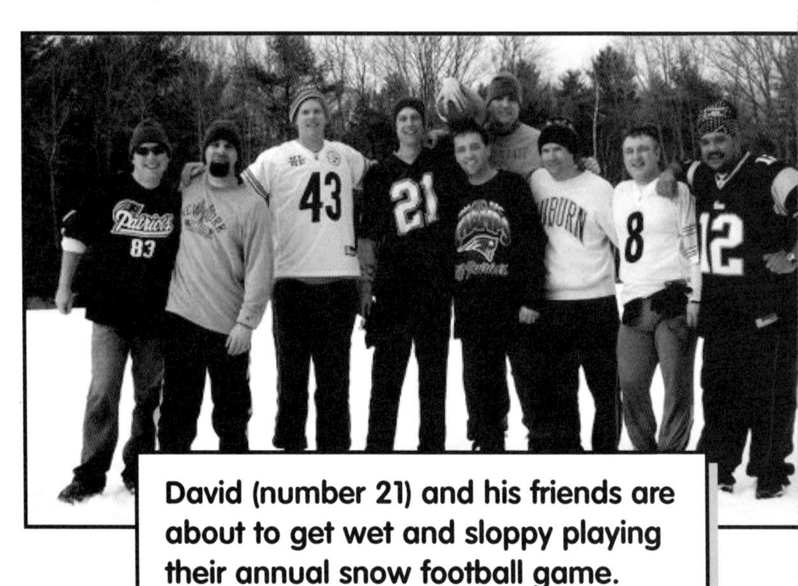

David (number 21) and his friends are about to get wet and sloppy playing their annual snow football game.

A statistician . . .

collects, analyzes, and presents numerical data that can explain past outcomes or predict future outcomes in everything from economics to politics. David uses statistics in sports to help predict how everything from individual plays to entire games might turn out. Other **statisticians**

- develop ways of predicting unemployment rates.
- examine voter surveys to forecast an election outcome.
- estimate wild salmon populations without counting individual fish.
- design experiments that compare two treatments for the same disease.

Is It 4 U?

As a sports statistician, David loves

- working with numbers.
- following the records of sports teams.
- helping people make decisions.
- watching both female and male athletes compete in sports.

What do you have in common with David? Write a paragraph describing three qualities you have that would make you a good sports statistician.

How Do You Measure Up?

Mathematicians track and study lots of different things. Track how much time you spend each day doing different things. Choose five things, such as doing homework, playing sports, practicing the piano, watching TV, or text messaging.

- Write down how many minutes you spend on each activity.
- Track your time for a week.
- Plot each activity on a bar or circle graph.

Do you notice any trends? Are there changes you'd like to make? Why? How would you make a change? Compare your chart with a partner. Is there something you could change together?

About You

David decided early in his life that he wanted to be really good at something. What would you like to excel in—a school subject, sports, music, theater? Write an informative paragraph in your About Me Journal. Describe what you'd like to be good at and what you can do to reach that goal.

Nancy Leveson

Massachusetts Institute of Technology

Thumbs Up

After studying math and management in school, Nancy Leveson worked at IBM but grew bored. So she traveled around the world for two full years. Her journey changed her. Meeting so many different types of people made her want to help others, and traveling alone gave her the confidence to go back to school to get a Ph.D.

Human Error

An aerospace company hired Nancy to make sure their torpedoes wouldn't turn around 180 degrees and hit the launchers. That's when she invented the field of software safety. Nancy later helped to design systems that prevent airplanes from colliding. Now she works less with computers and more with people. A spacecraft failure may result from a simple miscommunication between two groups of engineers. "Technical decisions are affected by social forces," Nancy says.

Safety First

When the Space Shuttle *Columbia* disintegrated in 2003, NASA called in Nancy for advice. She took a look at how engineers at NASA cooperated, or failed to, and told NASA what needed to change. Now she's on a team working to ensure the safety of future human spaceflight.

Nancy feels right at home when she's surrounded by aircraft.

A systems engineer . . .

works on large engineering projects. Nancy makes sure aerospace, nuclear power, medical, and transportation systems are safe and reliable. Other **systems engineers**

- plan new airports or highways.
- design new manufacturing systems.
- manage homeland security.
- make sure one computer system is compatible with another.

Mathelaughical

Q. Why did the algebra teacher confiscate a student's slingshot?

A. Because it was a weapon of math disruption.

Back and Forth

Imagine you are in charge of scheduling cargo flights between two airports. A cargo plane's crew can only work between 9 A.M. and 7 P.M. It takes a cargo plane 2 hours to fly from airport A to airport B, moving at an average speed of 300 kilometers (about 186 miles) per hour. On the return flight, the cargo plane has to fly into the wind, so it always takes 1 hour longer to fly back to airport A.

- How far apart are the two airports in kilometers? In miles?
- What is the average speed, in kilometers per hour, of the cargo plane when it's flying back to airport A from airport B?
- How many round-trip flights can you schedule one cargo plane to fly each day?

Changes for the Better

Nancy's travels around the world gave her the confidence to get her Ph.D. Write a paragraph about an experience that gave you the confidence to do something. Then share it with a partner. Discuss ways in which your experiences are similar and different.

Check out your answers on page 36.

About Me

The more you know about yourself, the better you'll be able to plan your future. Start an **About Me Journal** *so you can investigate your interests, and scout out your skills and strengths.*

Record the date in your journal. Then copy each of the 15 statements below, and write down your responses. Revisit your journal a few times a year to find out how you've changed and grown.

1. *These are things I'd like to do someday.*
   Choose from this list, or create your own.
   - Create software for animators
   - Design Earth-friendly buildings
   - Apply math to unsolved problems
   - Use math to choreograph professional dances
   - Figure out the best paths for spacecraft
   - Work on complex math problems
   - Use math to study treatments for diseases
   - Conduct cutting-edge research
   - Start a new company
   - Improve ways to use computers to solve problems

2. *These would be part of the perfect job.*
   Choose from this list, or create your own.
   - Working independently
   - Traveling
   - Teaching
   - Designing a project
   - Learning new things
   - Drawing
   - Solving problems
   - Managing money
   - Debating ideas
   - Writing

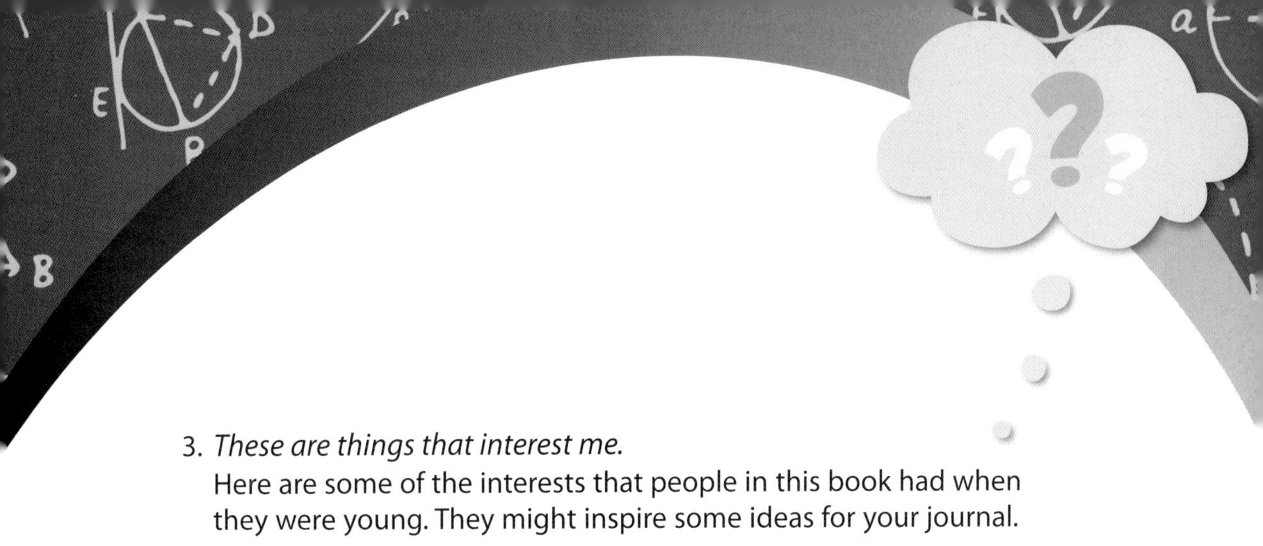

3. *These are things that interest me.*
   Here are some of the interests that people in this book had when they were young. They might inspire some ideas for your journal.
   - Protecting the environment
   - Playing sports
   - Solving math problems
   - Dancing
   - Joining a computer club
   - Going to math camp
   - Taking art classes
   - Making sculptures
   - Inventing things
   - Building new things out of old parts
   - Creating puzzles
4. *These are my favorite subjects in school.*
5. *These are my favorite places to go on field trips.*
6. *These are things I like to investigate in my free time.*
7. *When I work on teams, I like to do this kind of work.*
8. *When I work alone, I like to do this kind of work.*
9. *These are my strengths—in and out of school.*
10. *These things are important to me—in and out of school.*
11. *These are three activities I like to do.*
12. *These are three activities I don't like to do.*
13. *These are three people I admire.*
14. *If I could invite a special guest to school for the day, this is who I'd choose, and why.*
15. *This is my dream career.*

Careers 4 U!

Which Math career is 4 U?

What do you need to do to get there? Do some research and ask some questions. Then, take your ideas about your future—plus inspiration from scientists you've read about—and have a blast mapping out your goals.

On paper or poster board, map your plan. Draw three columns labeled **Middle School, High School,** and **College.** Then draw three rows labeled **Classes, Electives,** and **Other Activities.** Now, fill in your future.

Don't hold back—reach for the stars!

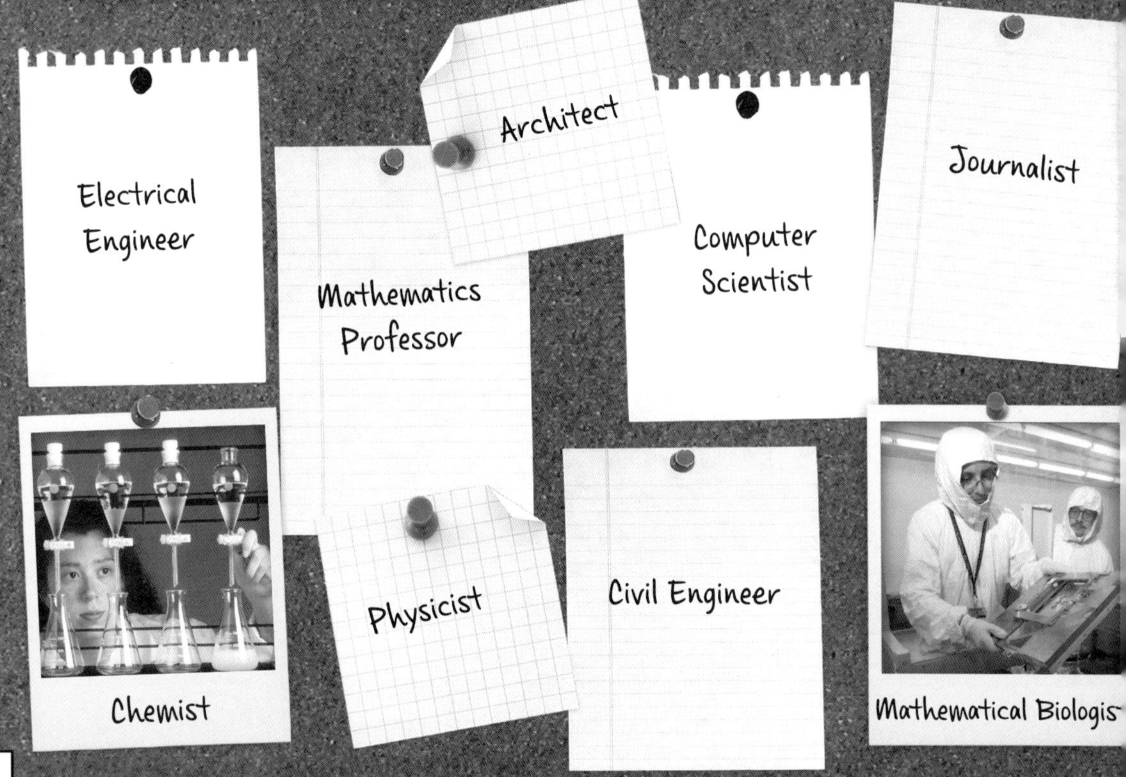

Sports Statistician

Cryptographer

Atmospheric Chemist

Biostatistician

Economist

Biologist

Financial Analyst

Systems Engineer

Epidemiologist

Mission Designer

Aeronautical Engineer

Physician

Robotics Engineer

Math Teacher

Choreographer

Planetary Scientist

Mathematician

algorithm (n.) A step-by-step procedure for solving a math problem, such as finding the greatest common divisor, or accomplishing some end especially by a computer. (p. 6)

apparent motion (n.) An optical illusion in which stationary objects viewed in quick succession or in relation to moving objects appear to be in motion. (p. 7)

computer code (n.) A set of instructions for a computer. (p. 14)

computer networks (n.) A collection of computers linked together so they can communicate with each other and share databases and peripheral devices. (p. 15)

computer model (n.) A simulation of some part of the natural world based on mathematical models. Simulations can run from minutes to hours to days, depending on the complexity of the system being studied. (pp. 14, 19)

conjecture (n.) A proposition, such as in mathematics, before it has been proved or disproved (p. 17)

constant (n.) A number that does not change or depend on other quantities. (p. 21)

database (n.) A large collection of information that is organized and stored in a computer. (p. 24)

engineering (n.) The application of science, math, and technology to design materials, structures, products, and systems. (pp. 12, 13, 29)

exponent (n.) A number that expresses how many times another number must be multiplied by itself. For example, in the expression x^n, n is the exponent of x. (p. 19)

forecast (v.) To calculate or predict some future event or condition, usually as a result of study and analysis of available information. (p. 27)

graph theory (n.) A branch of mathematics concerned with the study of graphs. (p. 16)

microchip (n.) A tiny complex of electronic components and their connections. Microchips are produced in or on a small slice of material, such as silicon. (p. 25)

probability (n.) A mathematical measure of how likely, or the chance that, an event will occur. If an event cannot happen, it is given a probability of 0, if it is certain, the probability is 1. (p. 14)

Punnett square (n.) A tool used in genetics to predict the probability of certain traits among offspring. It shows the different ways alleles can combine. An allele is one version of a particular gene. (p. 15)

Pythagoras (n.) A Greek philosopher and mathematician who lived from 560–480 B.C. and discovered that in a right-angled triangle, the length of the longest side, the hypotenuse, is related to the lengths of the other two sides by $c^2 = a^2 + b^2$ where *a*, *b*, and *c* are the lengths of the sides, with *c* being the hypotenuse. (p. 17)

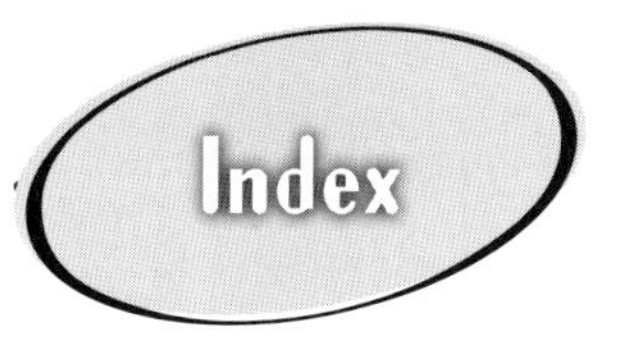

Index

CHECK OUT YOUR ANSWERS

ANIMATION SCIENTIST, page 7

Movie-Magic Math

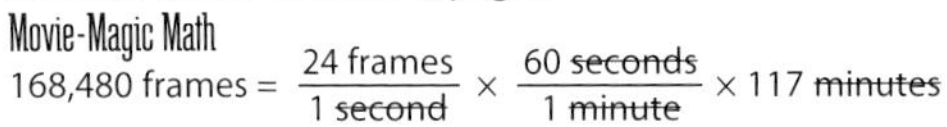

$168{,}480 \text{ frames} = \frac{24 \text{ frames}}{1 \cancel{\text{second}}} \times \frac{60 \cancel{\text{seconds}}}{1 \cancel{\text{minute}}} \times 117 \cancel{\text{minutes}}$

ARCHITECT, page 9

Geom-pet-ry *See Teacher Guide for calculations.*

The circumference is 103.62 meters.

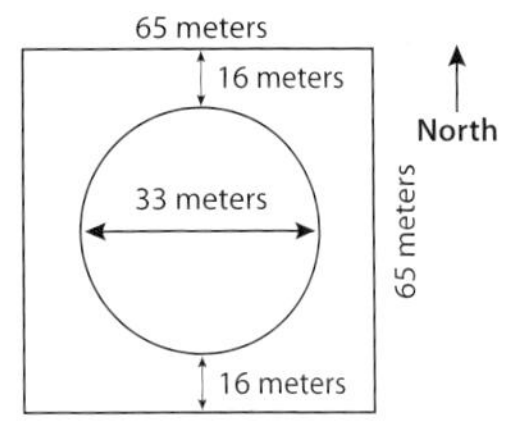

CIVIL SYSTEMS ENGINEER, page 13

Touch Down Time *See Teacher Guide for calculations.*

Flight 001 would arrive at 4 P.M. after a six-hour flight.
Flight 002 would arrive a 3 P.M. after a six-hour flight.
Flight 003 would arrive at 2 P.M. after a three-hour flight.

COMPUTER SCIENTIST, page 15

Odds on Dimples

	D	d
d	Dd	dd
d	Dd	dd

The probability of a child having dimples is 2/4 or 1/2, which equals 0.5 or 50%.

MATHEMATICAL BIOLOGIST, page 19

Why Cells R Little

Surface area formula 2(lw) + 2(hl) + 2(hw)

$\text{Ratio} = \frac{\text{Total surface area}}{\text{Total volume}}$

	1 4-centimeter cube	8 2-centimeter cubes	64 1-centimeter cubes
Total Surface area (square centimeters)	96	192	384
Total volume (cubic centimeters)	64	64	64
Surface area: volume	1.5:1	3:1	6:1

Growing Numbers

$12 \text{ five-minute intervals} = \frac{60 \text{ minutes}}{1 \cancel{\text{hour}}} \times \frac{1 \cancel{\text{hour}}}{5 \text{ minute intervals}}$

4,096 cells after hour = 2^{12} = 4,096 cells: 2 × 2 = 4 × 2 = 8 × 2 = 16 × 2 = 32 × 2 = 64 × 2 = 128 × 2 = 256 × 2 = 512 × 2 = 1,024 × 2 = 2,048 × 2 = 4,096

MATHEMATICIAN, page 21

Movie Math *See Teacher Guide for calculations.*

The average yearly change was $0.16 per year
The average price of a movie ticket in 2018 might be $8.78

A Penny Saved is a Penny Learned

$789 \text{ pennies} = 15 \cancel{\text{meters}} \times \frac{1 \text{ penny}}{1.9 \cancel{\text{centimeters}}} \times \frac{100 \cancel{\text{centimeters}}}{1 \cancel{\text{meter}}}$

$\$7.89 = 789 \cancel{\text{pennies}} \times \frac{\$1 \text{ dollar}}{100 \cancel{\text{pennies}}}$

The Winning Formulas *See Teacher Guide for calculations.*

1. Area of a triangle
2. Volume of a cone
3. Circumference of a circle
4. Pythagorean Theorem
5. Distributive Property of Multiplication
6. Commutative Property of Addition
7. Pi
8. Converting Celsius to Fahrenheit

MISSIONS DESIGNER, page 23

Sail Away *See Teacher Guide for calculations.*

Venus—22 minutes
Neptune—40 hours
Alpha Centauri—75 years

Hello Pluto

$2{,}087 \text{ million kilometers} = 13.95 \cancel{\text{AU}} \times \frac{149.6 \text{ million kilometers}}{1 \cancel{\text{AU}}}$

$1{,}297 \text{ million miles or } 1.30 \text{ billion miles} = 13.95 \cancel{\text{AU}} \times \frac{93 \text{ million miles}}{1 \cancel{\text{AU}}}$

NETWORKS EXPERT, page 25

Tiny Brains *See Teacher Guide for calculations.*

1. The number of transistors has doubled 20 times over the last 40 years.
2. Computer chips today can contain about 2 billion transistors.
3. Today's computer chips are about 1 million times more powerful than those 40 years ago.

It Makes Sense *See Teacher Guide for calculations.*

The sensors will cover 121,410 square meters which is equal to about 1,306,800 square feet.
Each sensor will cover 1,214.1 square meters which is equal to about 13,068 square feet.

SYSTEMS ENGINEER, page 29

Back and Forth *See Teacher Guide for calculations.*

The airports are 600 kilometers apart which is equal to about 372 miles.
The average speed is 200 kilometers per hour.
One cargo plane can be scheduled for two roundtrips each day.

Sally Ride Science is committed to minimizing its environmental impact by using ecologically sound practices. Let's all do our part to create a healthier planet.

These pages are printed on paper made with 100% recycled fiber, 50% post-consumer waste, bleached without chlorine, and manufactured using 100% renewable energy.